Z.B. TUCKER

# REAL Robots

Fun and Engaging Robot and Robot Age Lessons for Kids 5-16

# REAL Robots

**Robots** have often been the stuff of science fiction fantasy for decades. They have fascinated our imaginations with all the incredible possibilities that they could hold for humankind.

**Today** robots are more than just fantasy, they are "real" and exist among us, slowly changing the very way that we live.

**Each** day robots are becoming more essential to our society and the world is growing ever more dependent upon them to perform complex tasks and jobs. This book will explain the new era that we are currently living in. Ready or not, welcome to the age of **Real Robots!**

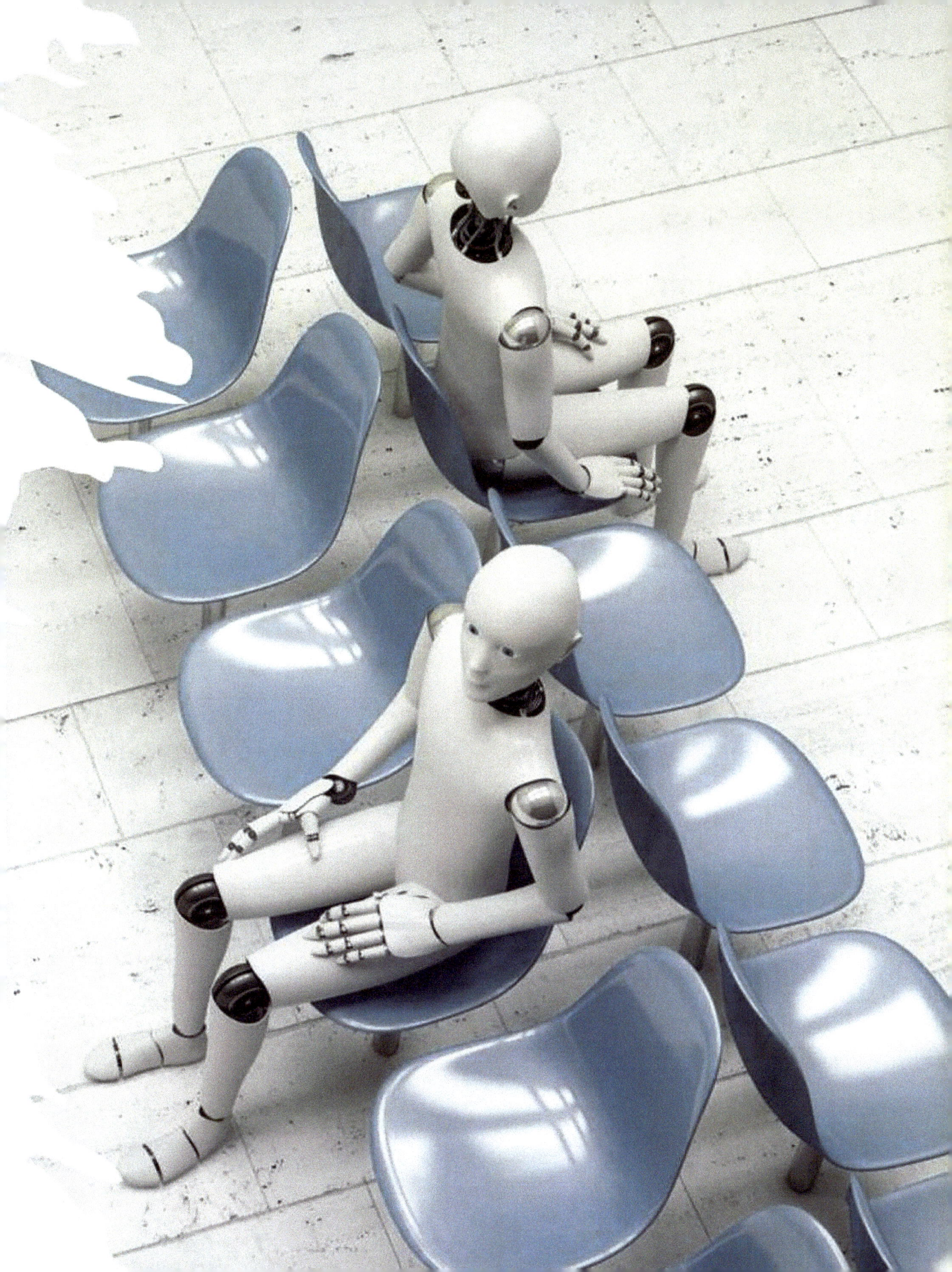

# Table of Contents

**Chapters** | **Page**

# 1 Robot Dedication

To all the budding scientist and their enthusiastic queries. To the engineers and visionaries of the future. To every child and their innate curiosity. This book is dedicated with great love and admiration to all who enjoy learning.

*Never stop learning*

# 2 Introduction to Robots

Let's define what a robot is and their purpose in the world.

# 3 What is a Robot?

A robot is a **machine**, especially one **programed** by a computer, capable of carrying out a complex series of actions automatically.

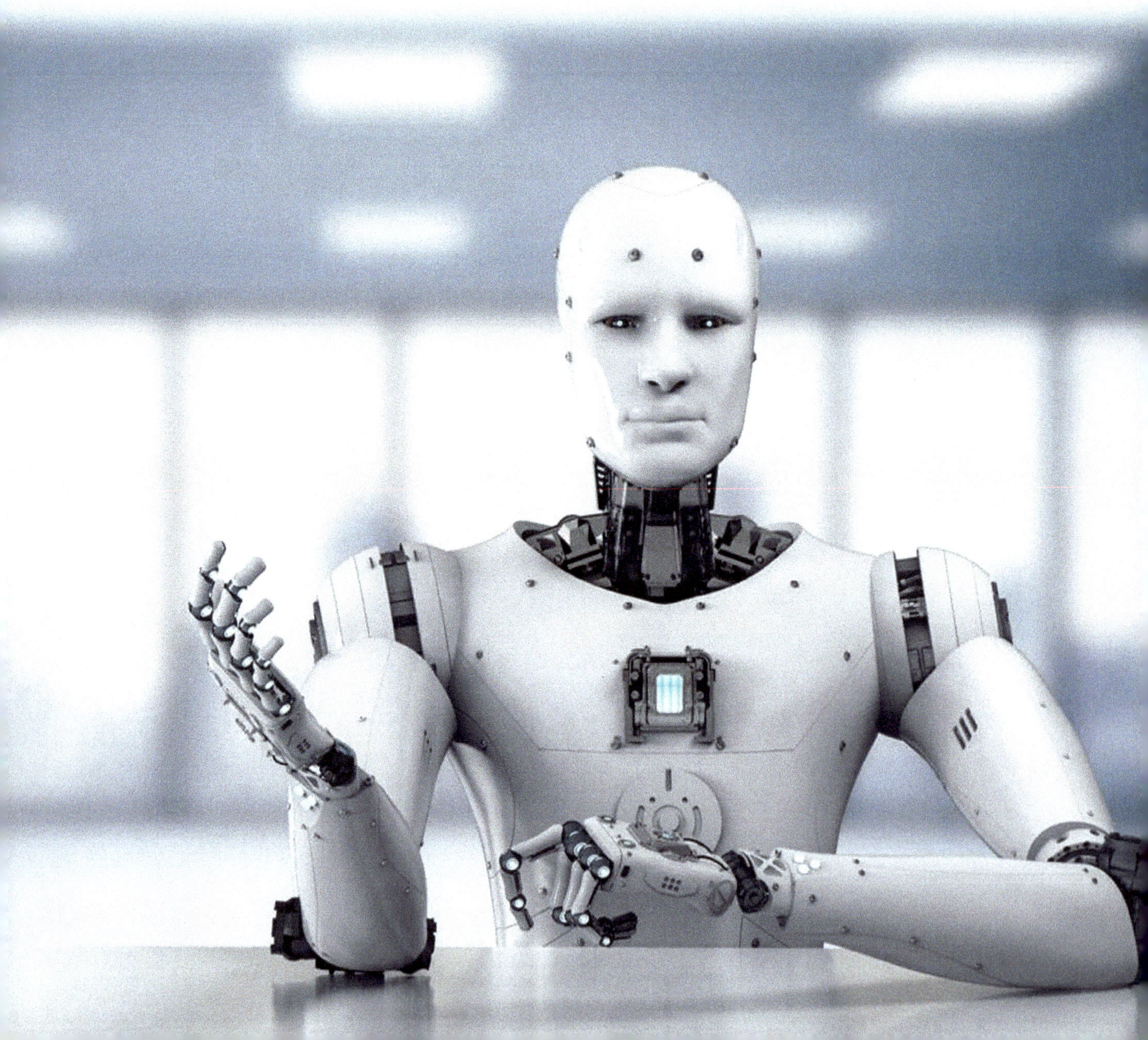

# 4 What is a Robot?

A robot can be guided by an **external** control device, or the control may be **placed** inside the robot.

# 5 What is a Robot?

Robots may be constructed to look **human**, but most robots are **task-performing** machines, designed with an emphasis on completing very specific set of actions.

# 6 What is the Purpose of a Robot?

**Robots** were originally built to handle **monotonous tasks** but have since expanded well beyond their initial uses. Each robot has a differing level of autonomy, ranging from human-controlled bots carrying out tasks that a human has full control over to fully-autonomous bots that perform tasks **without any external influence**.

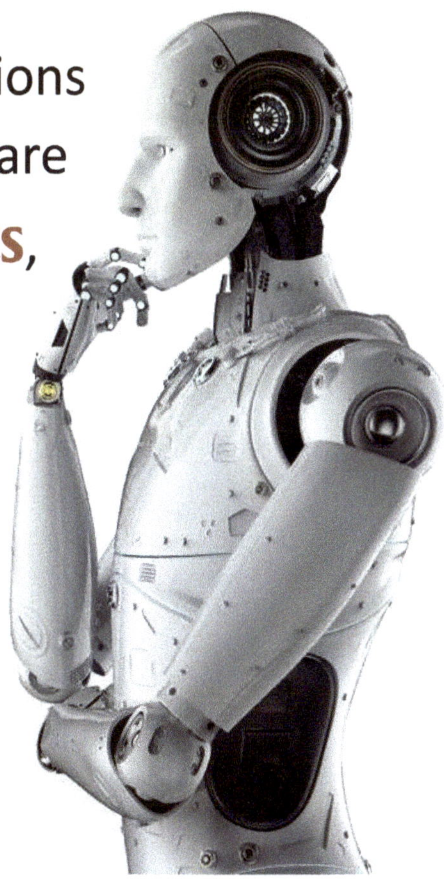

**Robots** can be used in many situations for many purposes, but today many are used in **dangerous environments**, including inspection of radioactive materials, bomb detection and deactivation, manufacturing processes, or where humans cannot survive (e.g., in space, underwater, in high heat, and clean up and containment of hazardous materials and radiation).

# 7 Brief Robot History

Let's explore how robots have developed over the years.

# 8 When Did It All Begin?

The concept of a robot originated many years ago throughout different cultures around the world. Early robots were often called Automatons. An **automaton** is a relatively self-operating machine, designed to automatically follow a sequence of operations, or respond to predetermined instructions.

Some of the earliest accounts of the idea of automatons date to the *4th century BCE* in Greek mythologies and various religious and philosophical texts from China. Engineers and inventors from ancient civilizations, including Ancient China, Ancient Greece, and Ancient Egypt, attempted to build self-operating machines, some resembling animals and humans. Automatons that resemble humans are called Humanoids.

# 9 Robot Historical Timeline

**1066** - In 1066, the Chinese inventor Su Song built a **water clock** in the form of a tower which featured mechanical figurines which chimed the hours throughout the day.

**1495** - Leonardo da Vinci an Italian engineer sketched plans for a **humanoid** robot that could perform basic movements like turn its head and wave its arms.

**1738** - Jacques de Vaucanson a French engineer created the **Digesting Duck** and the **Flute Player**. The Digesting Duck was able to imitate eating and flap its wings. It could even poop. The Flute Player was a wooden, human-sized robot that could play various melodies with the flute. It consisted of a system of bellows, pipes, weights, and other mechanical components to simulate what human muscles would have to do when playing the flute.

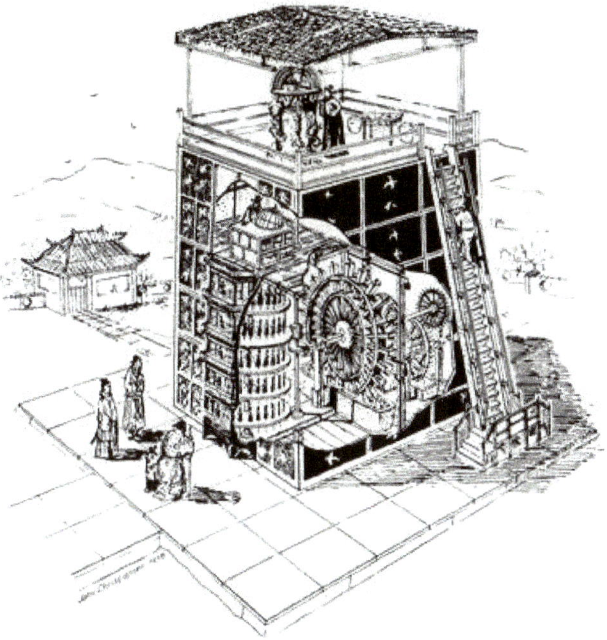

The Water Clock

The Leonardo Robot

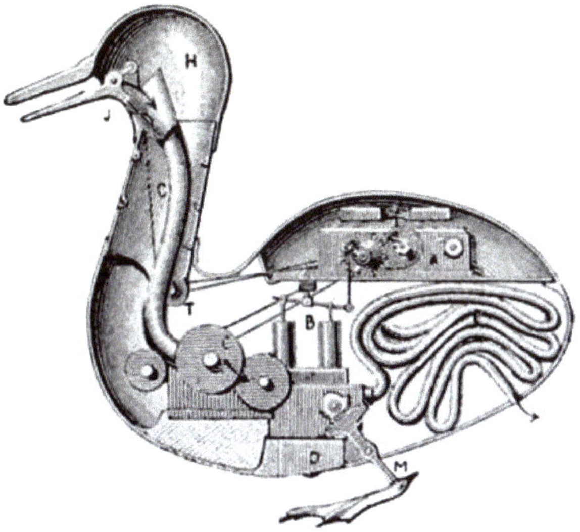

Digesting Duck

The Flute Player

# 10 Robot Historical Timeline

1796 - In Japan, complex animal and human automatons were built between the 17th to the 19th centuries. One such automaton was the **Karakuri Ningyō**, a mechanized puppet.

1928 - The **Eric robot** was the first British robot, built in 1928. Eric was constructed to open the Exhibition of the Society of Model Engineers. At the event's opening, Eric rose to his feet, bowed and gave a four-minute opening presentation.

1939 - **Elektro** was the nickname given to a robot built by the Westinghouse Electric Corporation in its Mansfield, Ohio facility. It was seven feet tall weighing 265 pounds and humanoid in appearance. Electro could walk by voice command, speak about 700 words, smoke cigarettes, blow up balloons, and move his head and arms.

Karakuri Ningyo

Electro Robot

Eric Robot

# 11 Robot Historical Timeline

1948 -The first electronic **autonomous** robots with complex behavior were created by *William Grey Walter* in Bristol, England in 1948. They were called **Tortoise Robots**.

1961 - **Unimate** was the first **industrial robot**, which worked on a General Motors assembly line at the Inland Fisher Guide Plant in Ewing Township, New Jersey.

1966 - **Shakey** was the world's first robot with **artificial intelligence**. It could perceive its surroundings, make plans, execute the plans, and correct its own errors. Many programming **algorithms** (instructions) used today descend from the Shakey Robot.

PLATE 53. *A mechanical "tortoise" which has been fitted with a type of computing circuit imitating a conditioned reflex.* **See page 156.**

Tortoise Robot

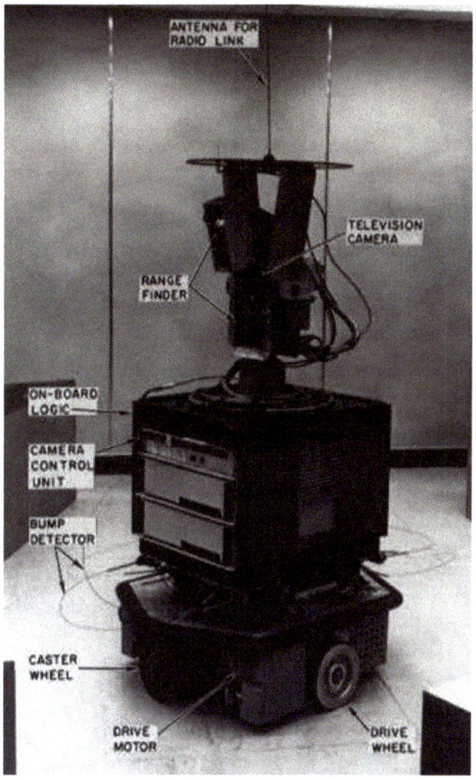


Shakey Robot

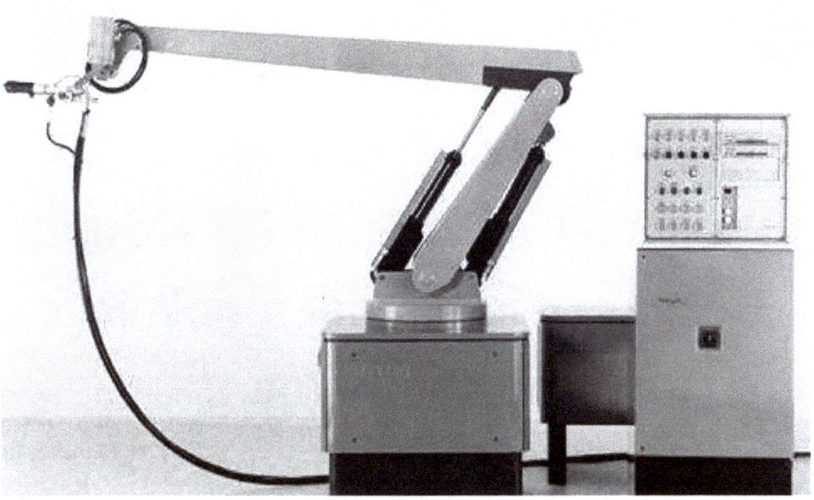

Unimate Robot

# 12 Robot Historical Timeline

1973 – **Wabot-1** First full-scale humanoid intelligent robot, and first **android (a robot that resembles a human)**. Its limb control system allowed it to walk, and to grip and transport objects with hands, using tactile sensors. Its vision system allowed Wabot-1 to measure distances and directions to objects using external receptors, artificial eyes and ears. And its conversation system allowed it to communicate with a person in Japanese, with an artificial mouth. He was constructed at Waseda University in Japan.

Today – According to the World Robotics 2020 Industrial Robots Report, a record of 2.7 million industrial robots are now operating in factories around the world. The highest density of industrial robots in the world is concentrated in Europe and America. Although **the country with the most robots per 1000 workers is Japan**.

Wabot-1

# 13 Parts of a Robot

Let's look at the five major components that make up a robot.

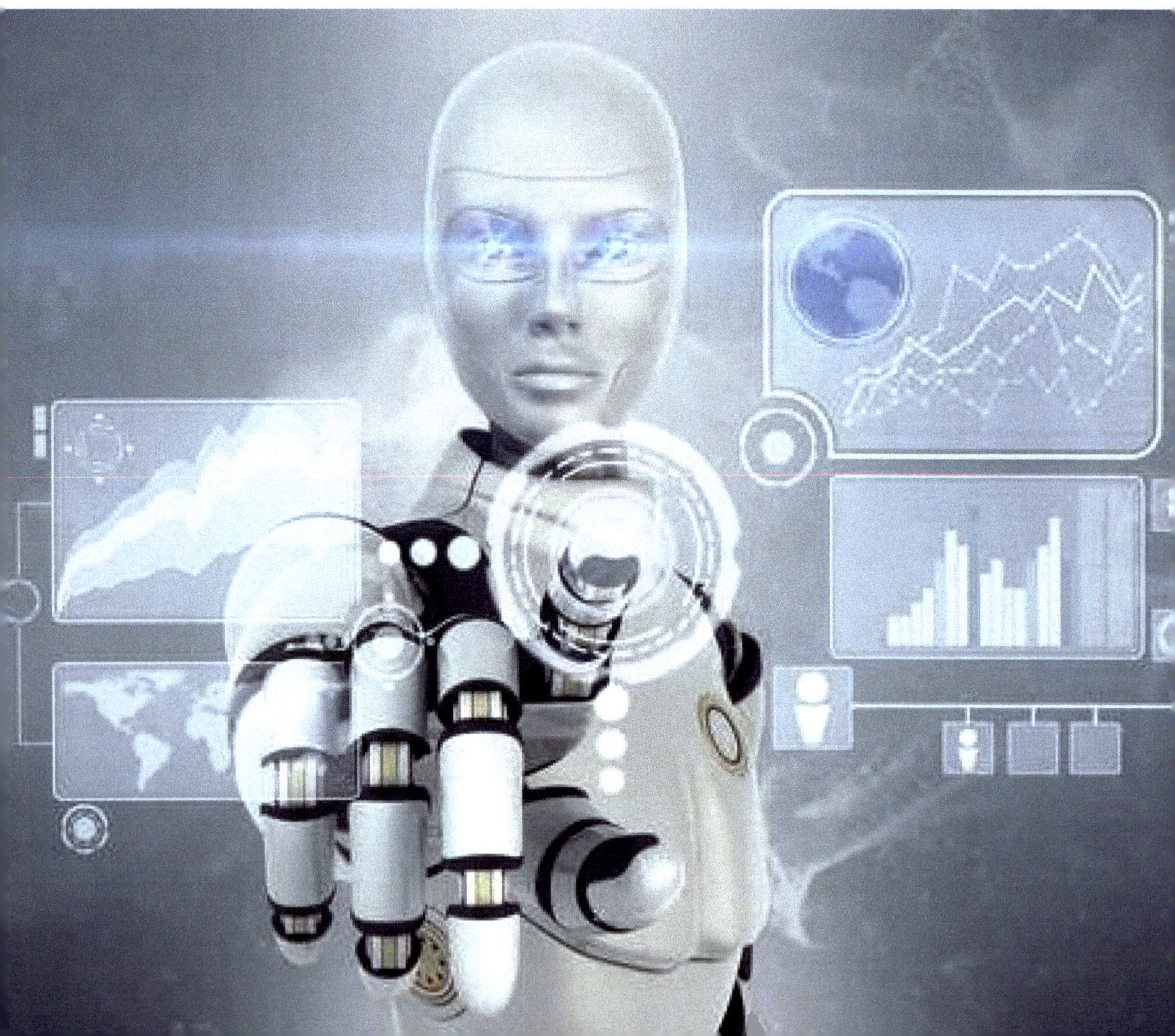

# 14 Parts of a Robot-1

1. **Body Structure** - Most robots consist of a **metal-based** physical body structure that provides form and protects the internal mechanisms within the robot. The body shape or structure is based on the robot's intended use or function. The most used metals for robotic body structures are **steel**, **copper**, and **brass**.

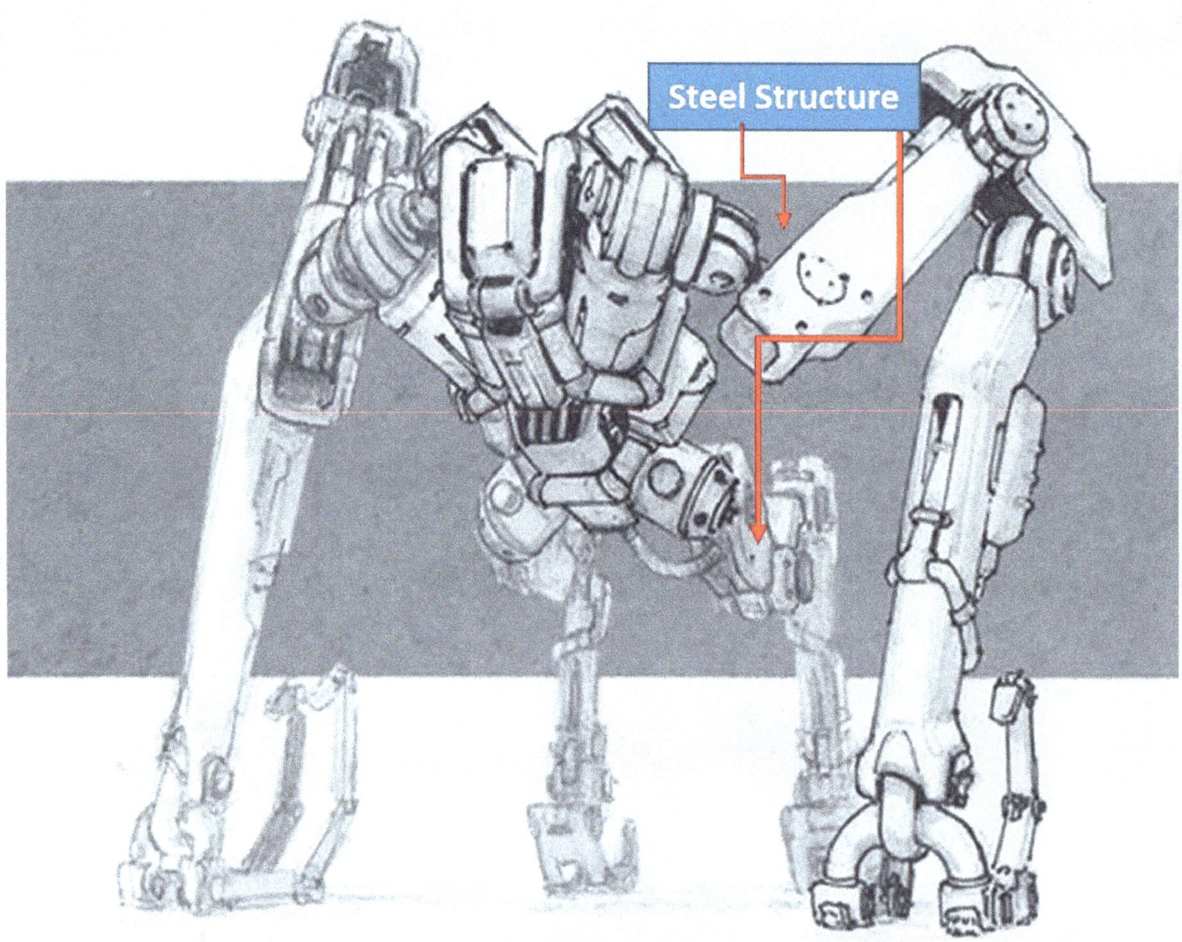

# 15 Parts of a Robot-2

**2. Power Source** - Like the human body requires food in order to function, robots require power. Stationary robots, such as those found in a factory, may run on **AC power** through a wall outlet but more commonly, robots operate via an **internal battery**.

# 16 Parts of a Robot-3

**3. Brain** – The Brain or controller of the robot is most often run by a **computer program**. A computer program gives the robot detailed commands and instructions to follow.

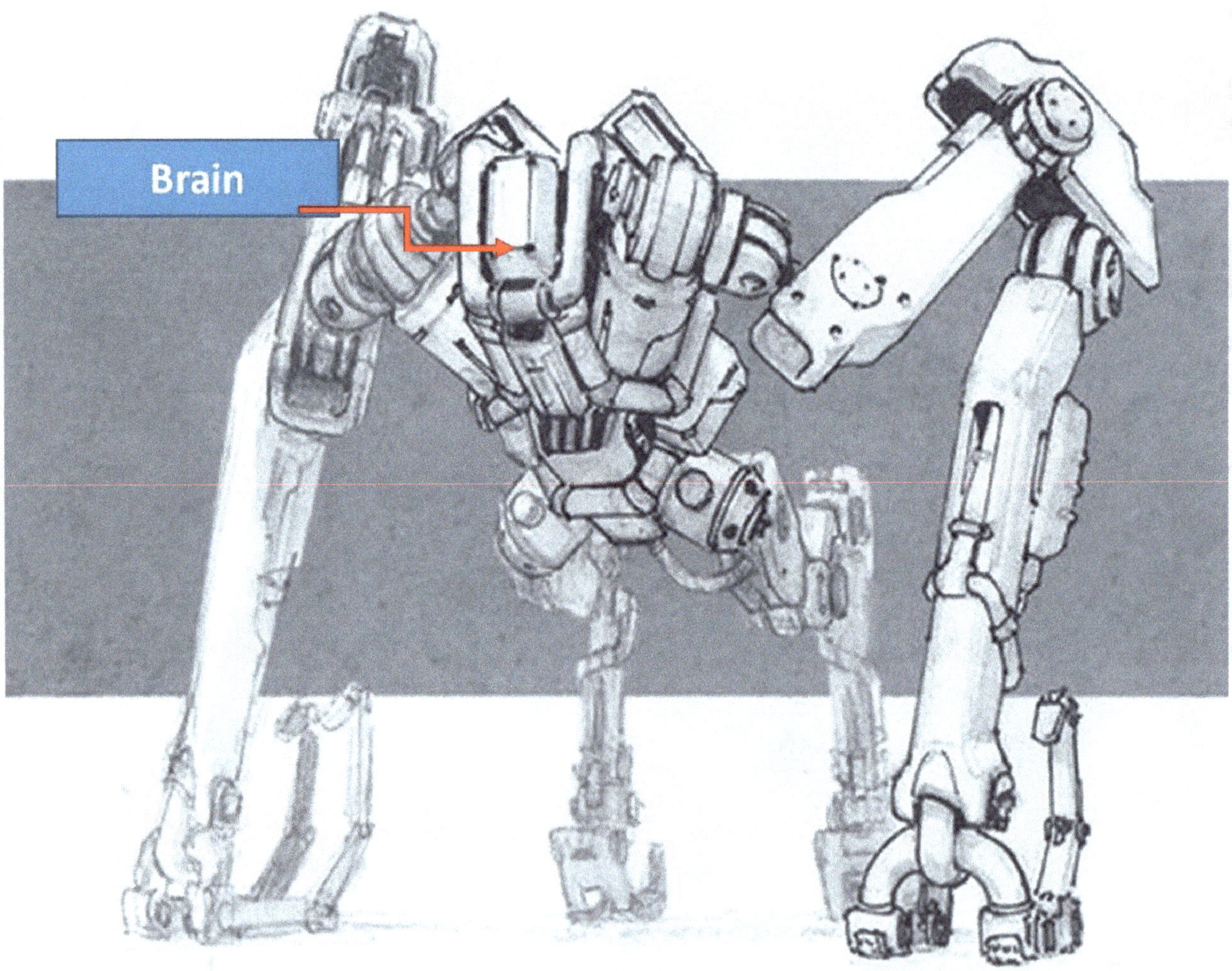

# 17 Parts of a Robot-4

**4. Mechanical Parts** – The mechanical parts include **motors, wheels, gears** and **grippers** to help the robot move, grab and turn. They can be powered by air, water and electricity.

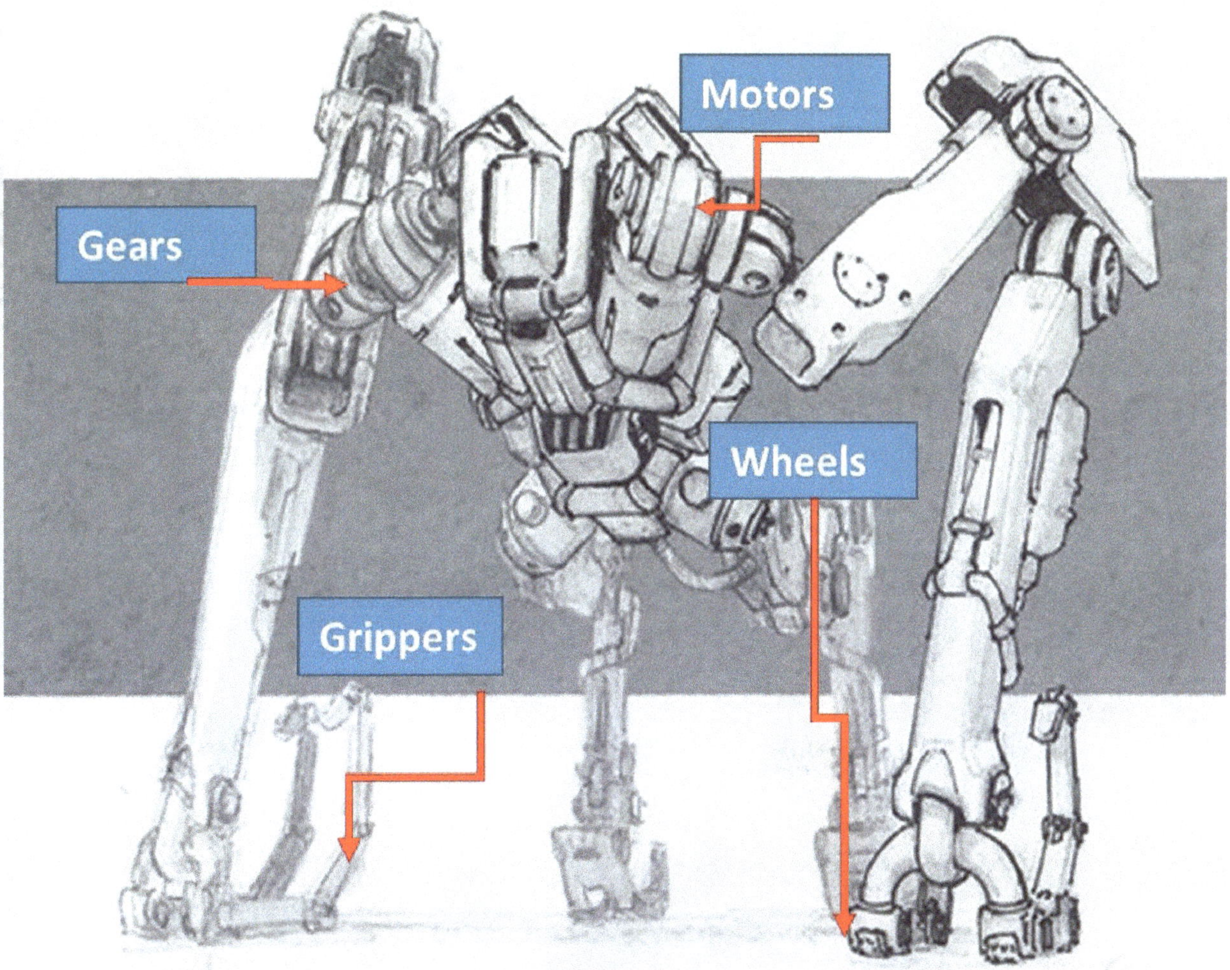

# 18 Parts of a Robot-5

5. **Sensors** – Sensors help robots determines **sizes, shapes, distance,** and **space** between objects

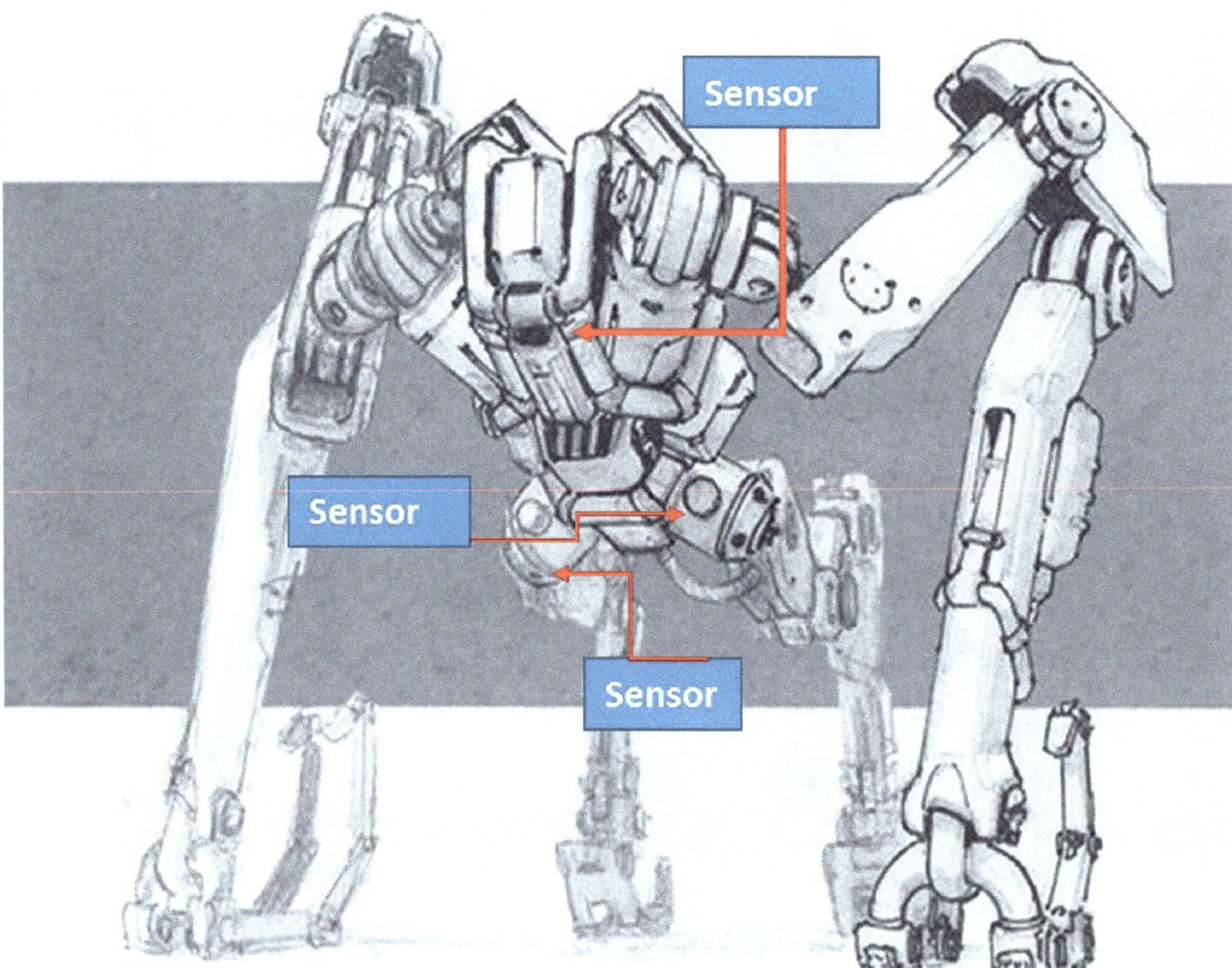

# 19 How do Robots Work?

# 20 Robot Sensors

Sensors provide the robot with information of its surrounding by perceiving different things. These sensors will transfer that information to the robot's brain for instructions on how to react to what it has detected. Sensors can detect many different sources of information such as:

- Light
- Sound
- Touch
- Smell
- Proximity
- And many other things...

# 21 Robot Brain

Once the sensors send information to the brain, also called the **Central Processing Unit (CPU)**, it will begin to search for pre-programmed instructions on how it should react to the information. These programmed instructions create a form of **computerized artificial intelligence (AI)**. AI allows a robot to perceive conditions and decide upon a course of action based on those conditions.

# 22 Mechanical Actions

Once the brain decides on the proper course of actions, instructions will be sent to the mechanical components of the robot and enable it to physically perform and carry out those activities.

# 23 Where Are Robots Used?

Let's discuss some of the places where robots are used in our society.

# 24 Where Are Robots Used?

**Car production** - Over the last three decades, automobile factories have become dominated by robots. A typical factory contains hundreds of industrial robots working on fully automated production lines, with one robot for every ten human workers. On an automated production line, a vehicle chassis on a conveyor is welded, glued, painted, and finally assembled at different sequences of robot stations.

# 25 Where Are Robots Used?

**Packaging** - Industrial robots are also used extensively for palletizing and packaging of manufactured goods, A palletizer is a machine that provides automatic means for stacking cases of goods or products onto a pallet. An example is a robot that can move drink cartons from the end of a conveyor belt and place them into boxes, or for loading and unloading goods at machining centers.

# 26 Where Are Robots Used?

**Electronics** - Mass-produced electronics are almost exclusively manufactured by pick-and-place robots, typically with manipulators, which remove tiny electronic components from strips or trays and place them onto printed circuit boards with great accuracy. Such robots can place hundreds of thousands of components per hour, far out-performing a human in speed and accuracy.

# 27 Where Are Robots Used?

**Space exploration** – Almost every unmanned space probe ever launched was a robot. Some were launched in the 1960s with very limited features, but their ability to fly and land (in the case of Luna 9) is an indication of their status as a robot. This includes the Voyager probes and the Galileo probes, among others.

# Where Are Robots Used?

**Travel** - Is there anything more science-fiction-like than autonomous vehicles? These self-driving cars are no longer just for sci-fi. A combination of data, science and robotics, self-driving vehicles are taking the world by storm. Automakers, like Tesla, Ford, Waymo, Volkswagen and BMW are all working on the next wave of travel that will let us sit back, relax and enjoy the ride. Rideshare companies Uber and Lyft are also developing autonomous rideshare vehicles that don't require humans to operate the vehicle.

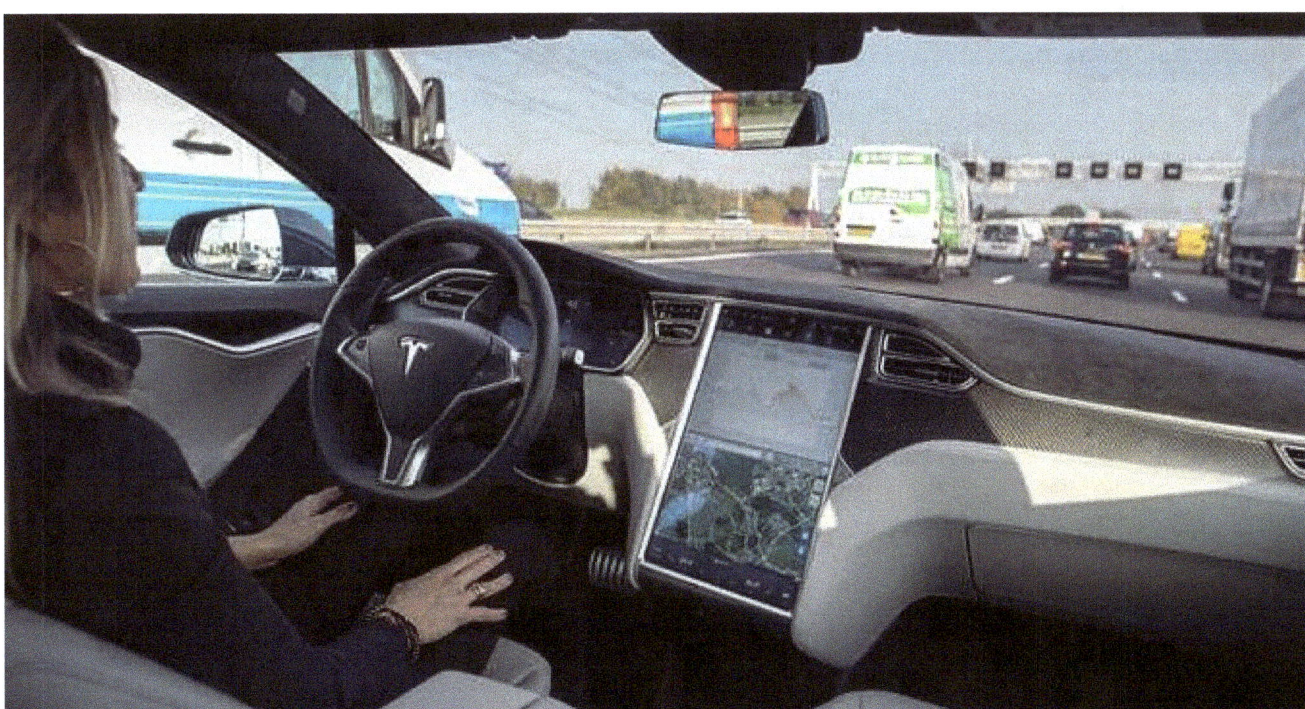

# 29 Where Are Robots Used?

**Healthcare** - Robots have made enormous strides in the healthcare industry. These mechanical marvels are being used in just about every aspect of healthcare, from robot-assisted surgeries to bots that help humans recover from injury in physical therapy. Examples of robots at work in healthcare is "**TUG**," a robot designed to autonomously stroll throughout a hospital and deliver everything from medicines to clean linens.

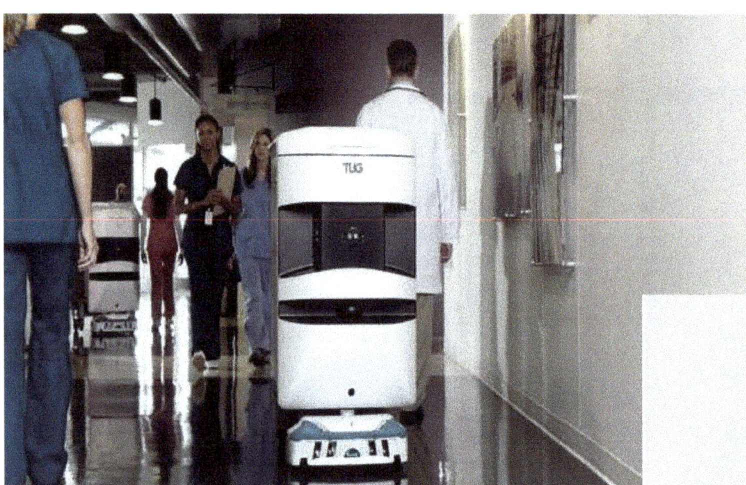


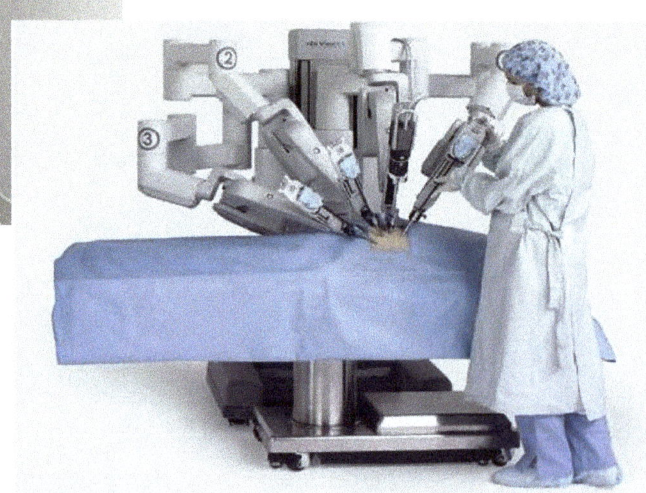

# 30 Robot Types

Robots can be classified into different categories. Let's review some of these categories.

# 31 Mobile Robots

**Mobile Robots** - have the capability to move around in their environment and are not fixed to one physical location. An example of a mobile robot that is in common use today is the *automated guided vehicle* (AGV). An AGV is a mobile robot that follows markers or wires on the floor or uses vision or lasers to move around

# 32 Industrial Robots

**Industrial Robots** - usually consist of a jointed arm (multi-linked manipulator) and an end effector that is attached to a fixed surface. One of the most common types of end effectors is a gripper assembly.

# 33 Educational Robots

**Educational Robots** – These robots are used as educational assistants to teachers. From the 1980s, such robot's called turtles were used in schools and programmed using a specific programming language called the Logo language.

# 34 Modular Robots

**Modular** – A Modular (self-reconfiguring robotic system) robot can change its own structural shape by rearranging their individual parts, in order to adapt to new circumstances, perform new tasks, or recover from damage. For example, a robot made of such components could assume a worm-like shape to move through a narrow pipe, reassemble into something with spider-like legs to cross uneven terrain, it can also be used for making “fixed” objects, such as walls, shelters, or buildings.

# 35 Collaborative Robots

**Collaborative** - A **cobot**, or **collaborative robot**, is a robot intended for direct human-robot interaction within a shared space, or where humans and robots are in close proximity. Cobot applications contrast with traditional industrial robot applications in which robots are isolated from human contact.

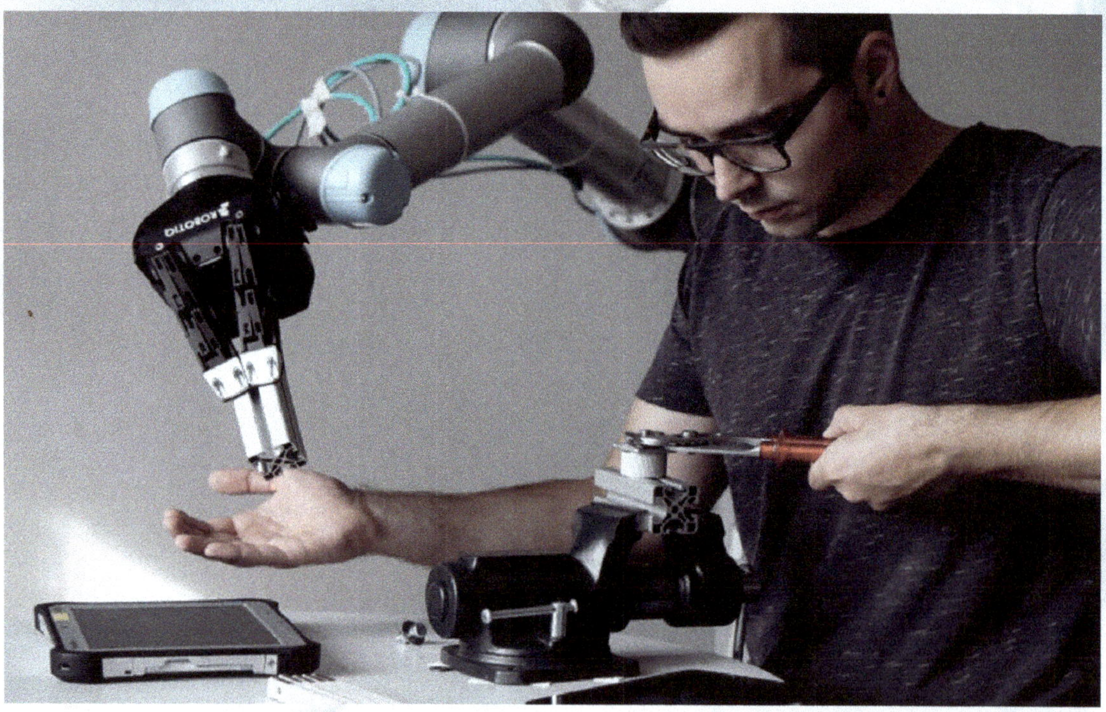

# 36 Humanoid Robots

**Humanoid** - A humanoid robot is a robot resembling the human body in shape. The design may be for functional purposes, such as interacting with humans, the environment, or for experimental purposes, such as the study of bipedal locomotion.

# 37 Ten of The World's Most Advanced Robots

**How advanced are robots today? Well, let's look at the top players...**

ROBOT
STOP
GO

**Pepper Robot** - Pepper is the world's first social humanoid robot able to recognize faces and basic human emotions. Pepper was designed for human interaction and can engage with people through conversation and his touch screen. Over 2,000 companies around the world have adopted Pepper as an assistant to welcome, inform and guide visitors in an innovative way.

**The Incredible Bionic Man -** Called Frank, short for Frankenstein, is a six-foot robot made up of artificial body parts donated by various research centers from around the world. Frank has 200 processors and is covered in over a million sensors.

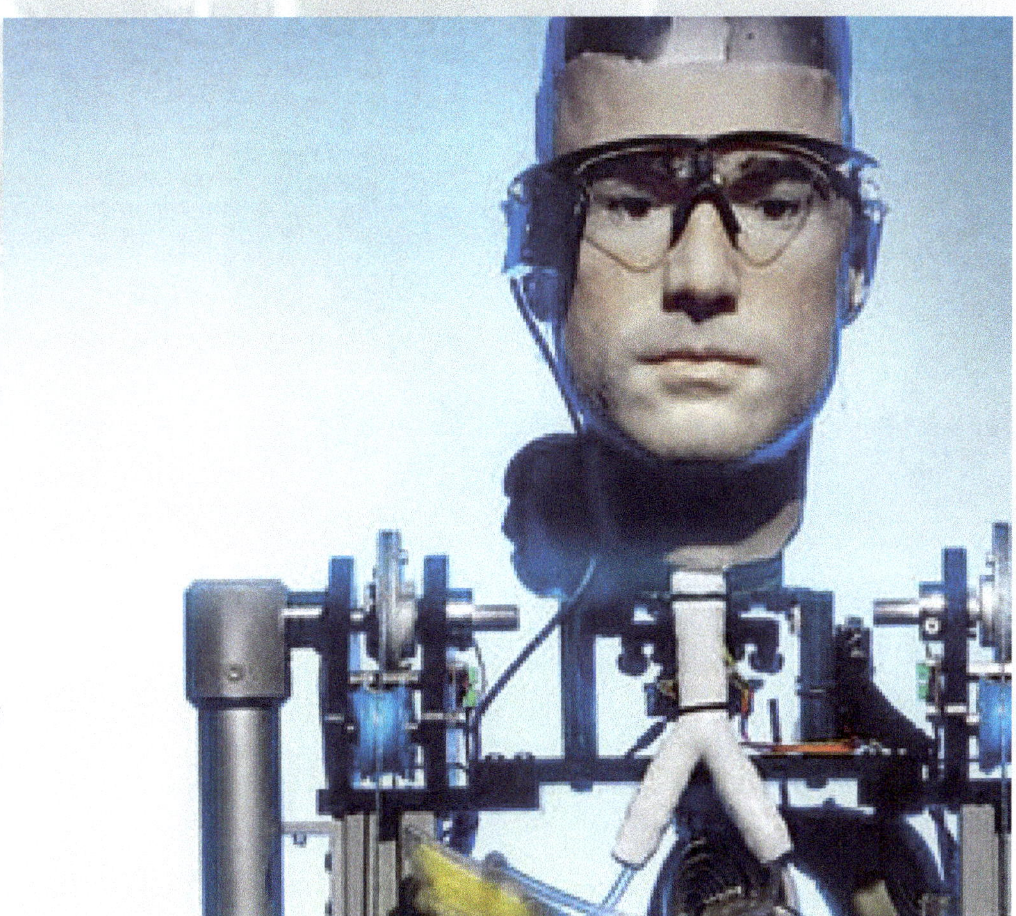

# 40 Most Advanced Robots

**Spot Mini** – Spot Mini is a compact, nimble four-legged robot that can trot around your office, home, or outdoors. It can map its environment, sense and avoid obstacles, climb stairs, and open doors. Spot can even fetch you a drink.

# 41 Most Advanced Robots

**Romeo** - Romeo is a 140 cm tall humanoid robot designed to explore and further research into assisting elderly people and those who are losing their autonomy. Romeo is a collaboration between numerous French and European laboratories and institutions. His size was determined to enable him to open doors, climb stairs and reach objects on a table.

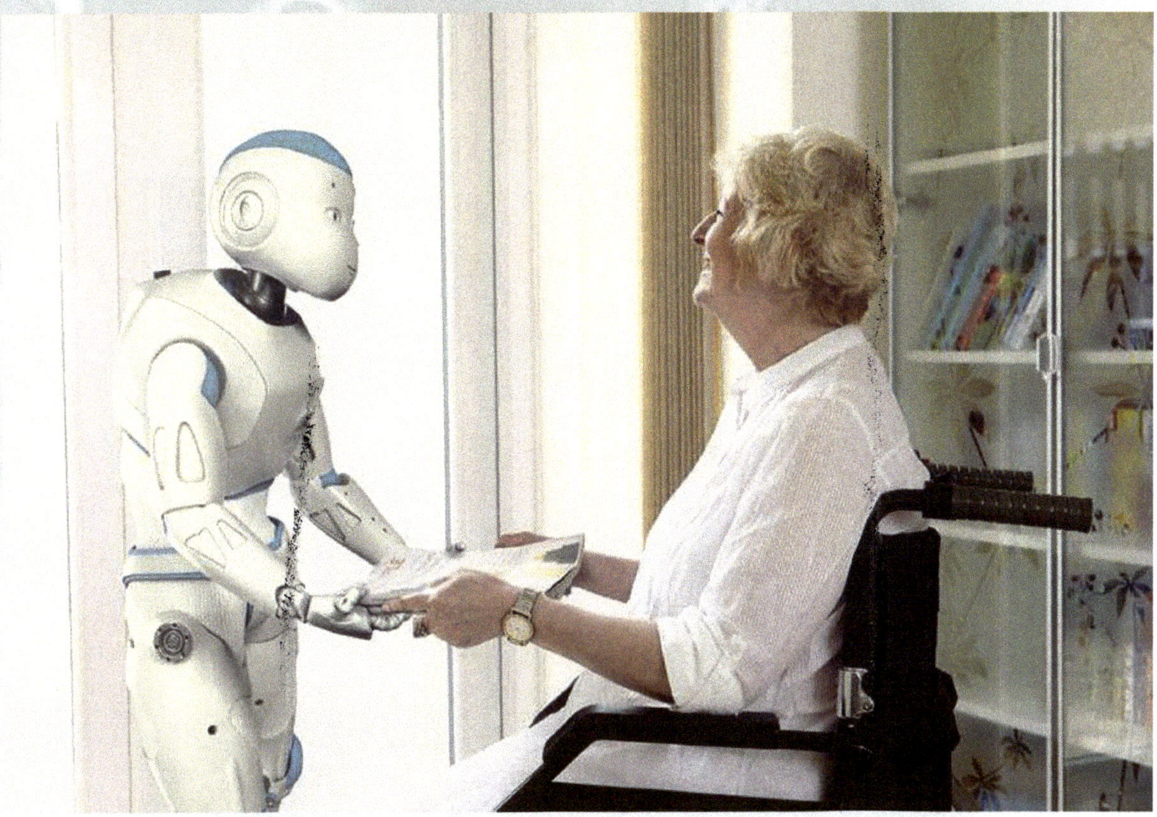

**Valkyrie Robonaut 5 NASA's (R5)** - is designed to be a robust, rugged, entirely electric humanoid robot capable of operating in extreme degraded or damaged environments. It is an ideal robot for space exploration and new planet colonization like Mars.

# 43 Most Advanced Robots

**Sawyer** - Is the world's most advanced industrial robot. Sawyer can perform over 150 task that are either too difficult or too repetitive for humans to perform.

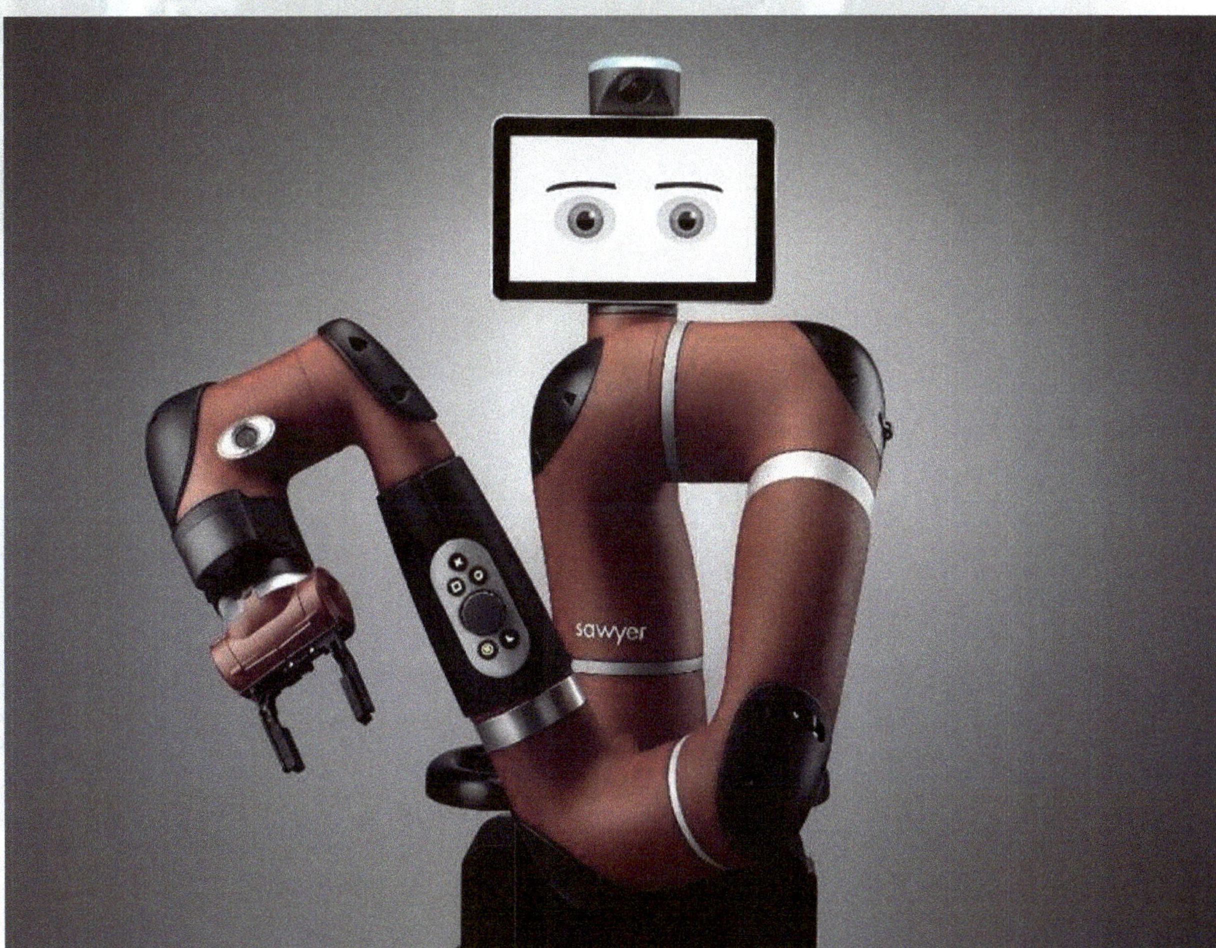

**Method 2** – A robot that can be driven from the inside to perform a wide variety of tasks. Method 2 is designed to protect people from extremely dangerous areas like fires.

# 45 Most Advanced Robots

**Atlas** – Created by the company Boston Dynamics with capabilities that allow it to run 1.5 meters per second. He is a very agile and versatile robot designed for highly precise athletic movements including jumps and backflips. Atlas is also a very good dancer.

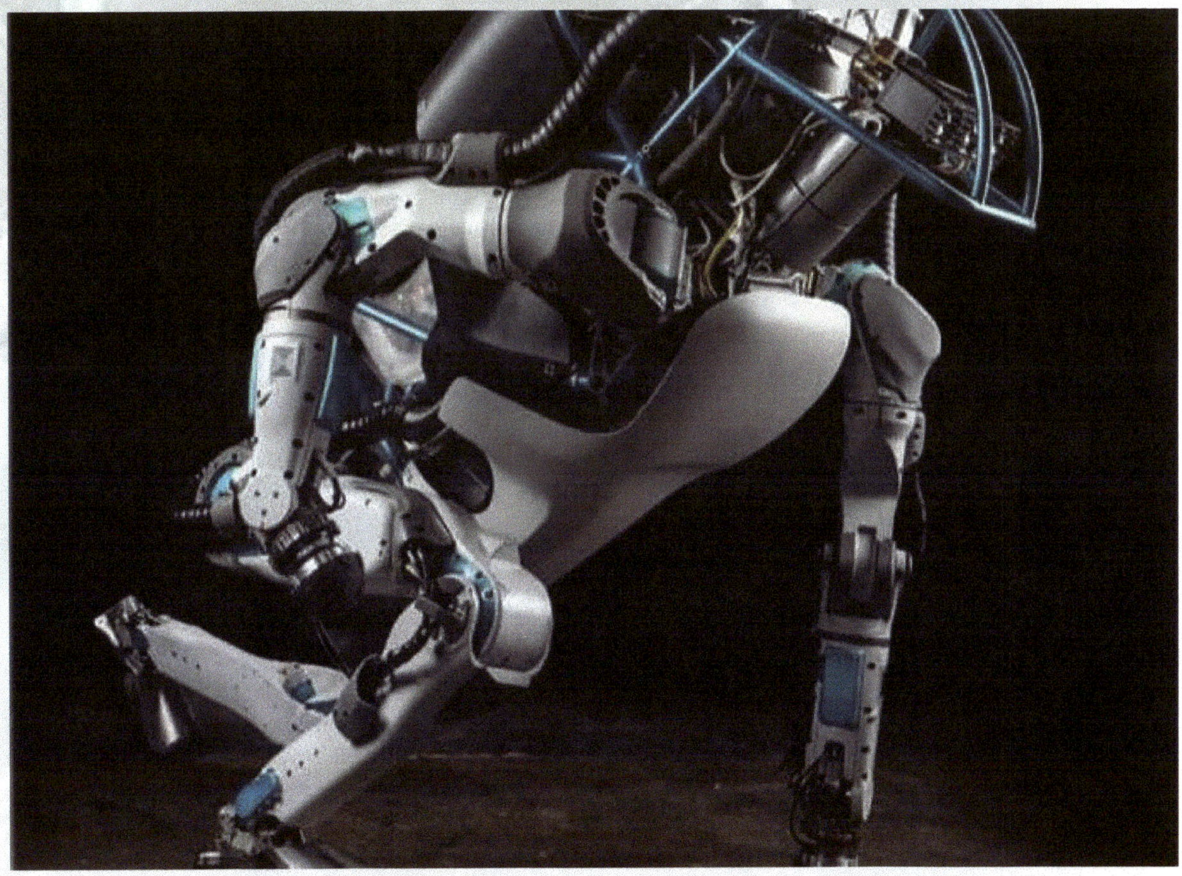

**ASIMO** – Designed by Honda Manufacturers, ASIMO was engineered to assist people with limited abilities in a wide variety of helpful ways. Among other things, ASIMO can serve coffee and use sign language to speak.

# 47 Most Advanced Robots

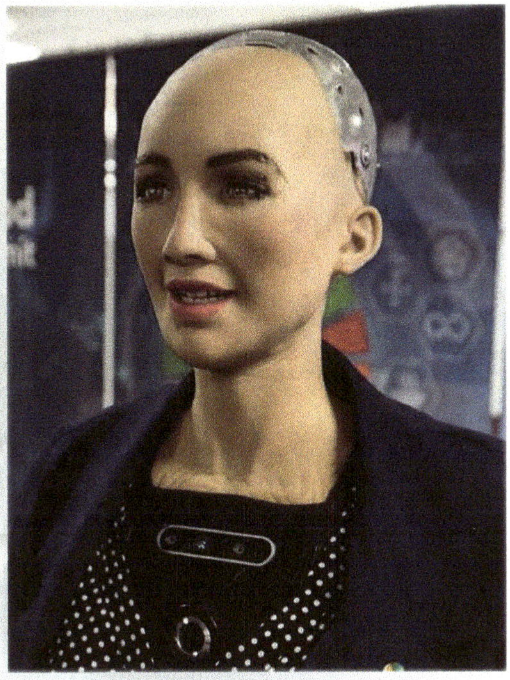

**Sophia** – Sophia is a highly sophisticated social humanoid robot developed by the Hong Kong-based company Hanson Robotics. Sophia was activated on February 14, 2016. She made her first public appearance in mid-March 2016 at South by Southwest (SXSW) in Austin, Texas, United States.

Sophia has been covered by media around the globe and has participated in many high-profile interviews. In October 2017, Sophia was given Saudi Arabian citizenship and became the first robot to receive citizenship of any country. In November 2017, Sophia was named the United Nations Development Programme's first Innovation Champion and is the first non-human to be given a United Nations title.

# 48 The Future of Robots

Let's look at the areas where robots will play a larger role in the coming years.

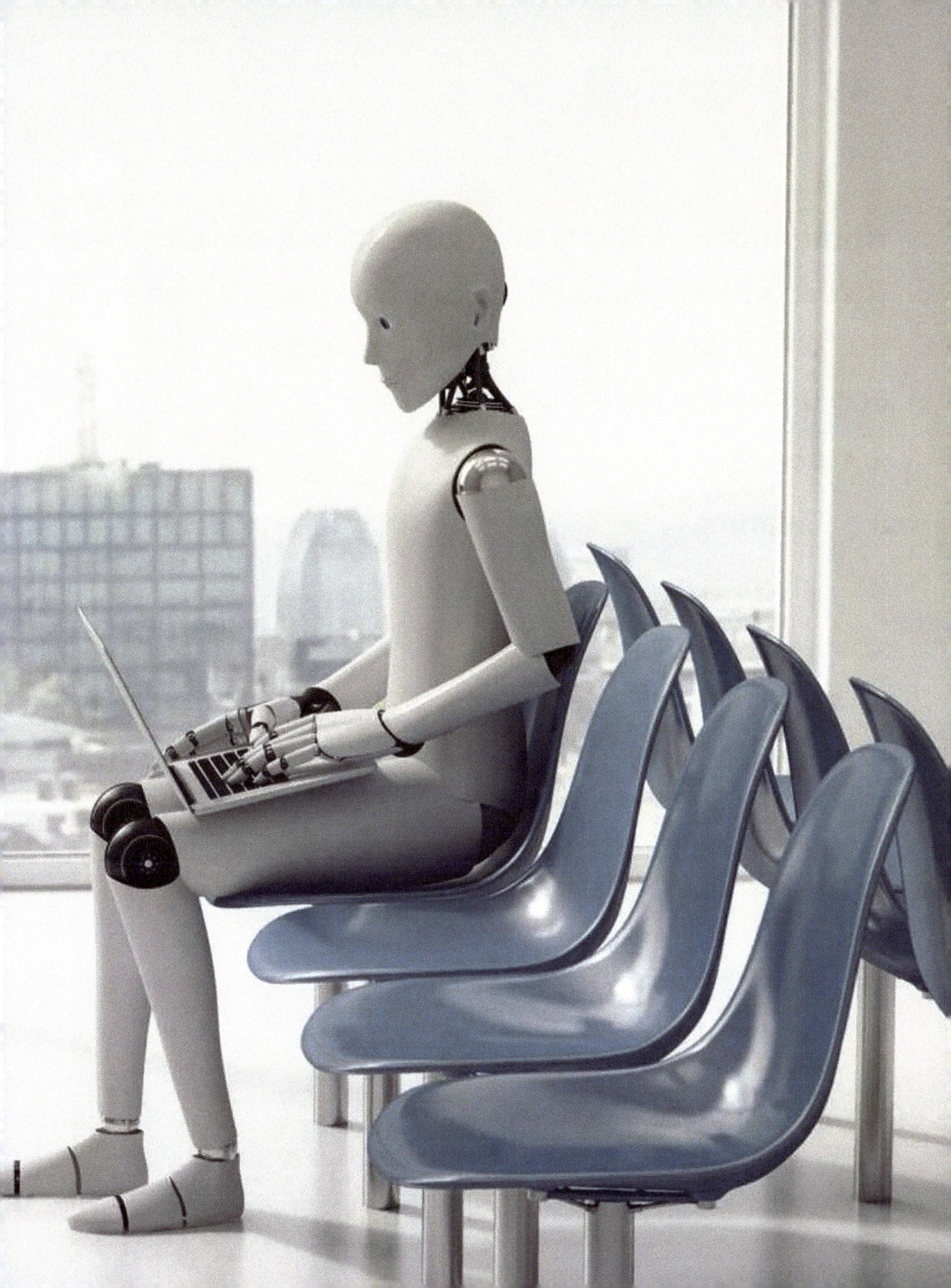

# 49 Home Robots

**Home robots** – This is one of the fastest-growing areas that robots are expanding into. These service bots can perform a wide range of jobs both indoors and outdoors including:

- **Cooking**
- **Cleaning**
- **Lawncare**
- **Security**

# 50 Air Taxis

**An Air Taxi** - is a flight charter service, typically operating between small airports, to provide people with on-demand flights to locations within the taxi's service region. This is a growing area and will one day be flown automatically without any onboard human pilots.

# 51 Emotional Robots

**Emotional robots** – help with loneliness and education, teach languages and help with exercise. This is another growing field that will expand in the coming future.

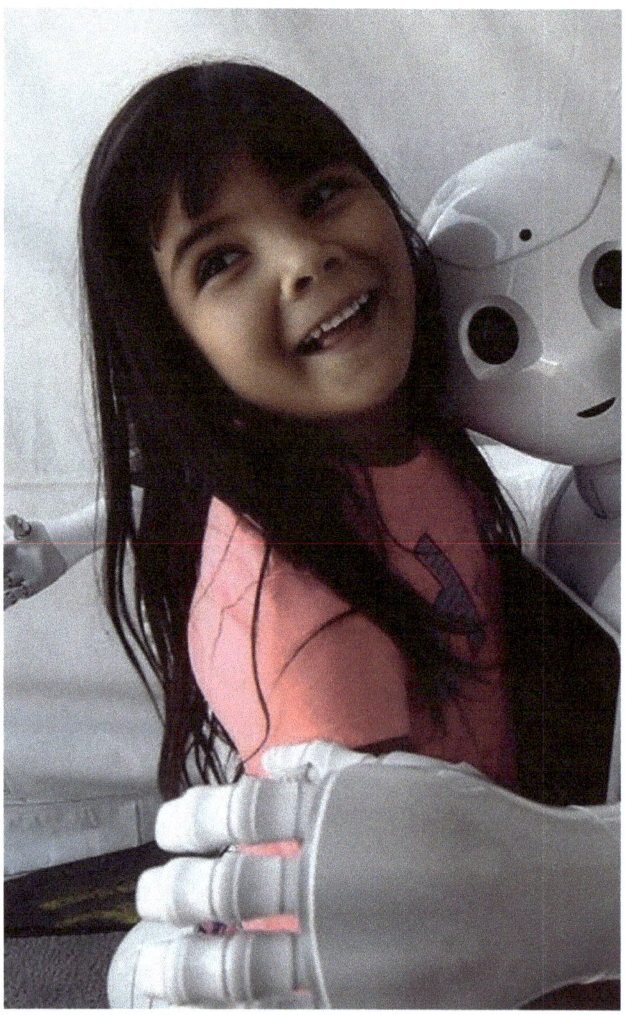

# 52 Military Robots

**Military robots** - are autonomous robots or remote-controlled mobile robots designed for military applications, from transport to search & rescue and attack. Some such systems are currently in use, and many more are under development.

# 53 Agricultural Robots

An agricultural robot, also known as an **agribot**, is a robot designed for use in the agriculture industry. These agribots help farmers with harvesting, planting, pruning, sorting and picking.

# 54 Can Robots Learn On Their Own?

If you are wondering if robots can learn new things beyond the limits of their programming, the answer is YES! This is called Machine learning (ML).

Machine Learning is the study of **computer algorithms** (programmed instructions) that can improve a robot's learning ability automatically through experience and data that it receives over time. It is seen as a part of a robot's artificial intelligence development.

Machine learning algorithms builds data, known as **training data**, that helps to make predictions and robotic decisions without being specifically programmed to do so.

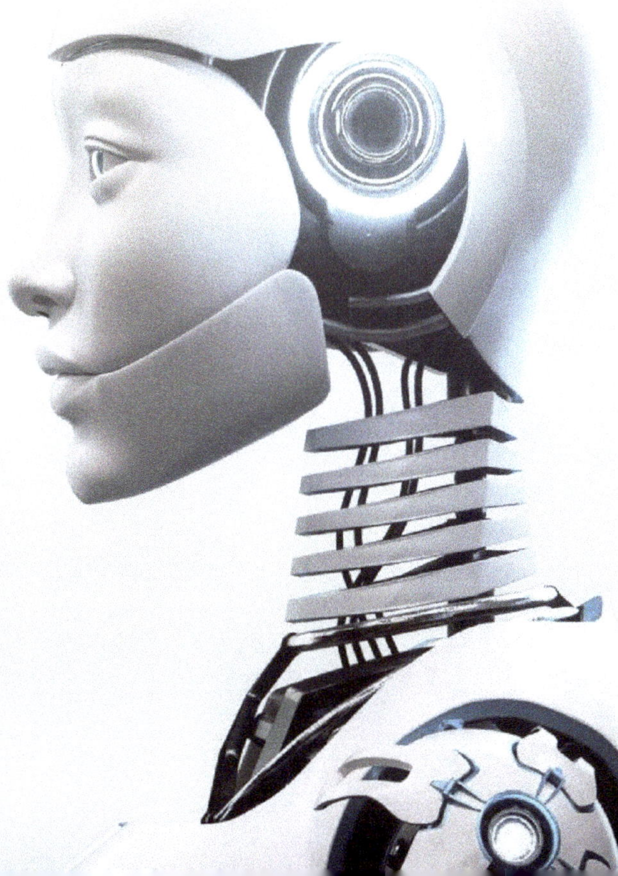

# 55 Build Your Own Robot

Want to build your own robot? Many stores sale robot kits to make robot building fun and easy to learn and understand for children. A good kit will give you a lot of the basic foundations for becoming a future **robot builder**, **programmer** or **designer**.

# 56 Robot POP Quiz

Let's review what all you have learned

# 57 Choose The Correct Answer

**Which parts make up a robot?**

| | | | |
|---|---|---|---|
| 1. | Body Structure | Yes | No |
| 2. | Hair | Yes | No |
| 3. | Brains | Yes | No |
| 4. | Ears | Yes | No |
| 5. | Sensors | Yes | No |
| 6. | Power source | Yes | No |
| 7. | Stomach | Yes | No |
| 8. | Mechanical parts | Yes | No |

# 58 Write a short answer:

- What is a robot?

- What is the purpose of a robot?

- What is an automaton?

59

# Fill in the blanks

1. ____________ is the name of the robot who can jump, do backflips and dance.

2. ____________ is the first robot to become an official citizen of Saudi Arabia.

3. An ___________ is a flight charter service, typically operating between small airports.

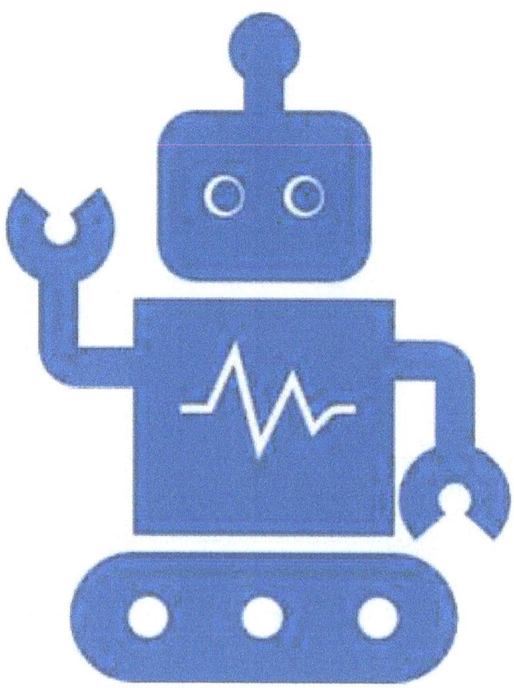

# 60 Robotic Definitions

1. **Algorithm** - An algorithm is a set of instructions for solving a problem or accomplishing a task.
2. **Artificial Intelligence** – is a term used for machines capable of perception, logic and learning.
3. **Android** – A robot with a human appearance.
4. **Automaton** – A moving mechanical device made to imitate human beings.
5. **Autonomous** – Having the freedom to operate without direct human control.
6. **Bot** – A robot
7. **Central Processing Unit** - The part of a computer in which operations are controlled and executed.
8. **Computer Program** - A computer program is a sequence of instructions, written to perform a specified task for a computer.
9. **End Effector** - The end effector is the part of the robot that interacts with the environment.
10. **Humanoid** - Having an appearance or character resembling that of a human.
11. **Robotics** - the branch of technology that deals with the design, construction, operation, and application of robots.

# 61 The End

Robots are real and play a very important role in our world today. They help us in many ways by performing difficult jobs and tasks. Their numbers will soon grow and expand, and they will become smarter and more capable as time goes on because they can learn new things. Our greatest hope and dream for the future is that robots will help humans create a better and brighter world for us all to live in.

# Other Books By The Author

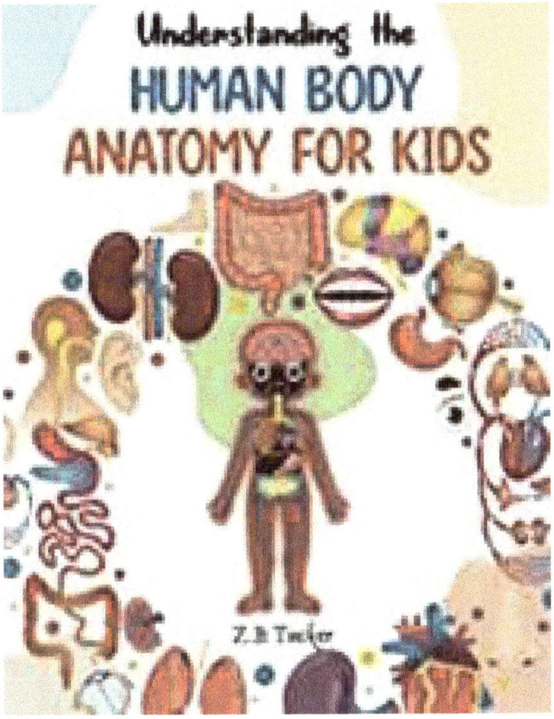

# 63 Author Bio

The author is an educator and a passionate lover of children. He received his Bachelor of Arts in Biology from Edgewood College in Madison, Wisconsin and a second B.A. in Healthcare Management from Southern Illinois University Carbondale where he also completed his M.A in Engineering. He has a wealth of valuable experience dealing with children of varying age ranges. The author does not only love children, but he understands that when they have the best form of education, it has a far-reaching effect on the wider society. As a result, he is on a mission to make sure that children learn as much as they can.

# References


Global Agriculture Robots Market Size By Type (Driverless Tractors, Automated Harvesting Machine, Others), By Application (Field Farming, Dairy Management, Indoor Farming, Others) By Geography Scope and Forecast (Report). Verified Market Research. September 2018.

Harris, Tom (16 April 2002). "How Robots Work". How Stuff Works. Archived from the original on 26 August 2007. Retrieved 10 September 2007. https://science.howstuffworks.com/robot.htm

Currie, Adam (1999). "The History of Robotics". Archived from the original on 18 July 2006. Retrieved 10 September 2007. https://web.archive.org/web/20060718024255/http://www.faculty.ucr.edu

Fowler, Charles B. (October 1967). "The Museum of Music: A History of Mechanical Instruments". Music Educators Journal.

Needham, Joseph (1991). Science and Civilization in China: Volume 2, History of Scientific Thought. Cambridge University Press.

Pearce, Jeremy. "George C. Devol, Inventor of Robot Arm, Dies at 99" Archived 25 December 2016 at the Wayback Machine, The New York Times, 15 August 2011. Retrieved 7 February 2012.

"Earliest Clocks". A Walk-Through Time. NIST Physics Laboratory. Archived from the original on 31 May 2008. Retrieved 11 August 2008.

Jane Marie Law, Puppets of Nostalgia – The Life, Death and Rebirth of the Japanese Awaji Ningyo Tradition, 1997, Princeton University Press.

Wood, Gabby. "Living Dolls: A Magical History Of The Quest For Mechanical Life" Archived 20 December 2016 at the Wayback Machine, The Guardian, 16 February 2002.

"AH Reffell & Eric Robot (1928)". Archived from the original on 11 November 2013. Retrieved 11 November 2013.

# 65 References

"AH Reffell & Eric Robot (1928)". Archived from the original on 11 November 2013. Retrieved 11 November 2013.

Owen Holland. "The Grey Walter Online Archive". Archived from the original on 9 October 2008. Retrieved 25 September 2008. https://web.archive.org/web/20081009055230/http://www.ias.uwe.ac.uk/Robots/gwonline/gwonline.html.

Scott Schaut (2006). Robots of Westinghouse: 1924-Today. Mansfield Memorial Museum.

"Robot Hall of Fame – Unimate". Carnegie Mellon University. Archived from the original on 26 September 2011. Retrieved 28 August 2008.

"History of Industrial Robots". robots.com. Archived from the original on 8 July 2015. Retrieved 24 August 2015.

"Robotics in practice: Future capabilities" by Joseph F. Engelberger. in "Electronic Servicing & Technology" magazine 1982 August.

Robots: Brave New World moves a step closer Archived 14 January 2019 at the Wayback Machine, By James Melik, reporter, Business Daily, BBC World Service, 3 January 2013.

"Kitchen robot in Riga cooks up new future for fast food". techxplore.com. Retrieved 14 August 2021.

IFR International World Report. https://ifr.org/ifr-press-releases/news/record-2.7-million-robots-work-in-factories-around-the-globe

Artificial Intelligence in Design '96. Springer, Dordrecht. pp. 151–170

www.ingramcontent.com/pod-product-compliance
Lightning Source LLC
LaVergne TN
LVHW070856210926
848482LV00016B/954
*9798997244996*